Collins
INTERNATIONAL PRIMARY SCIENCE

Workbook 5

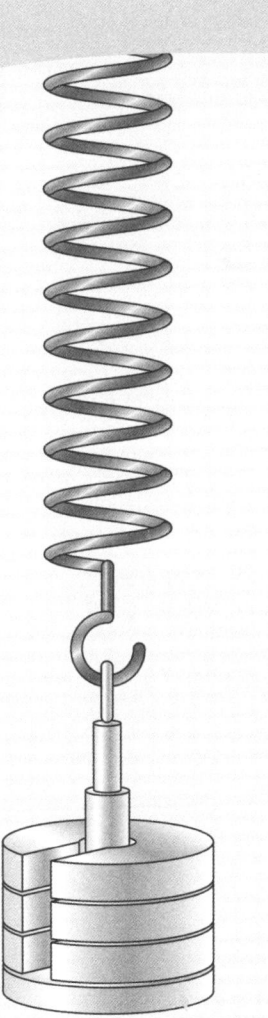
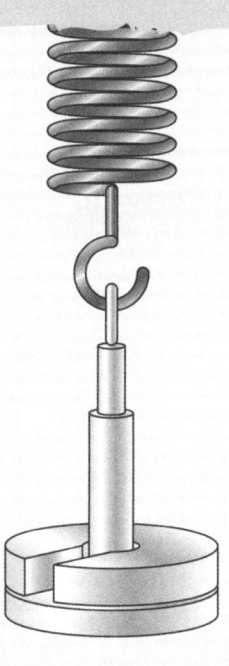
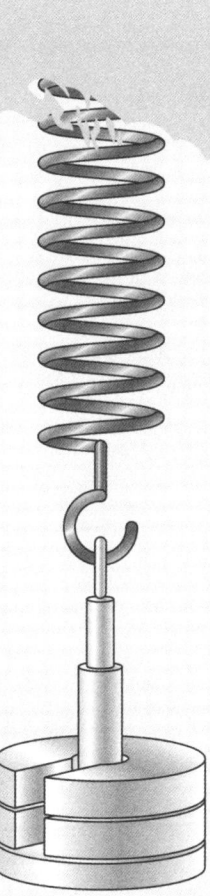

William Collins' dream of knowledge for all began with the publication of his first book in 1819. A self-educated mill worker, he not only enriched millions of lives, but also founded a flourishing publishing house. Today, staying true to this spirit, Collins books are packed with inspiration, innovation and practical expertise. They place you at the centre of a world of possibility and give you exactly what you need to explore it.

Collins. Freedom to teach.

Published by Collins
An imprint of HarperCollins*Publishers* Ltd.
The News Building
1 London Bridge Street
London
SE1 9GF

HarperCollins*Publishers*
Macken House,
39/40 Mayor Street Upper,
Dublin
D01 C9W8
Ireland

Browse the complete Collins catalogue at
www.collins.co.uk

© HarperCollins*Publishers* Limited 2021

10 9 8 7

ISBN: 978-0-00-836897-5

Second edition

This book is produced from independently certified FSC™ paper to ensure responsible forest management.

For more information visit:
www.harpercollins.co.uk/green

Contributing authors: Karen Morrison, Tracey Baxter, Sunetra Berry, Pat Dower, Helen Harden, Pauline Hannigan, Anita Loughrey, Emily Miller, Jonathan Miller, Anne Pilling, Pete Robinson.

All rights reserved. No part of this publication may be reproduced, stored in a retrieval system, or transmitted in any form or by any means, electronic, mechanical, photocopying, recording or otherwise, without the prior written permission of the Publisher or a licence permitting restricted copying in the United Kingdom issued by the Copyright Licensing Agency Ltd, 5th Floor, Shackleton House, 4 Battle Bridge Lane, London SE1 2HX

British Library Cataloguing in Publication Data
A Catalogue record for this publication is available from the British Library.

Commissioning editor: Joanna Ramsay
Product manager: Letitia Luff
Development editor: Karen Williams
Project manager: 2Hoots Publishing Services Ltd
Proofreader: Caroline Low
Cover designer: Gordon MacGilp
Cover illustrator: Ann Paganuzzi
Image researcher: Emily Hooton
Illustrators: Beehive Illustration (John Batten, Moreno Chiacchiera, Phil Garner, Kevin Hopgood, Tamara Joubert, Andrew Pagram, Simon Rumble, Jorge Santillan, Matt Ward)
Internal design and typesetting: Ken Vail Graphic Design Ltd
Production controller: Lyndsey Rogers
Printed in India by Multivista Global Pvt. Ltd.

With thanks to the following teachers and schools for reviewing materials in development: Preeti Roychoudhury, Sharmila Majumdar and Sujata Ahuja, Calcutta International School; Hawar International School; Melissa Brobst, International School Budapest; Rafaella Alexandrou, Diana Dajani, Sophia Ashiotou and Adrienne Enotiadou, Pascal Primary School Lefkosia; Niki Tzorzis, Pascal Primary School Lemesos; Vijayalakshmi Chillarige, Manthan International School; Taman Rama Intercultural School.

Acknowledgements
The publishers wish to thank the following for permission to reproduce photographs.
Every effort has been made to trace copyright holders and to obtain their permission for the use of copyright materials. The publishers will gladly receive any information enabling them to rectify any error or omission at the first opportunity.

p15a Vova Shevchuk/Shutterstock, p15b azure1/Shutterstock, p15c asadykov/Shutterstock, p15d Xray Computer/Shutterstock, p15e Scisetti Alfio/Shutterstock, p15f xpixel/Shutterstock, p24 Andrew Astbury/Shutterstock, p32l Serg64/Shutterstock, p32c & p33t l i g h t p o e t/Shutterstock, p32r Nature Capture Realfoto/Shutterstock, p33b terekhov igor/Shutterstock, p38 PhotoStock-Israel/Alamy Stock Photo, p52l Angyalosi Beata/Shutterstock, p52r Semmick Photo/Shutterstock.

Third-party websites, publications and resources referred to in this publication have not been endorsed by Cambridge Assessment International Education.

Registered Cambridge International Schools benefit from high-quality programmes, assessments and a wide range of support so that teachers can effectively deliver Cambridge Primary.

Visit www.cambridgeinternational.org/primary to find out more.

Contents

Topic 1 Our living world

Planning a fair test	1
The parts of plants (1)	2
The parts of plants (2)	3
The parts of a flower	4
Drawing a cross-section of a flower	5
What happens during fertilisation?	6
Investigating seeds	7
Ways that flowers attract insects (1)	8
Ways that flowers attract insects (2)	9
Insects and flowers	10
Ways that seeds can be dispersed and spread around	11
Animals that disperse seeds	12
Making a model of a spinning seed	13
What we found out about seeds with wings	14
Seed dispersal key	15
Conditions for germination	16
Seeds growing into plants	17
Stages in the life of a plant	18
The life cycle of a sunflower	19
A plant that is adapted to its environment	20
An animal in a cold climate	21
Predators and prey	24
Predators and prey relationships	25

Topic 2 Healthy eating

The organs of the human digestive system	26
How food moves through the digestive system	27
Food pyramid diagram	28
A balanced meal	29

Topic 3 Materials

Diagrams show what matter looks like	30
How matter behaves and changes state	31
States of matter diagram	32
A solid changes to a liquid	33
Puddle investigation	34
Drying clothes	35
Investigating condensation	37
Collecting water	38
Heating ice	39
What happens when you heat ice?	40
Does sugar dissolve?	42
Temperature and dissolving	43
Making crystals	44

Topic 4 Forces

Use a table to record observations	45
Using a line graph	46
What do you know about forces?	47
Investigate friction on different surfaces	48
Investigating friction	49
Parachute investigation	51
Investigate air resistance on cars	52
Does shape affect water resistance?	54
Design a toy which travels through water	55
Magnetic or non-magnetic?	56
Are the poles of a magnet equally strong?	57
Magnetic forces	58
Forces and distance	59

Topic 5 Sound

The string telephone	60
Plastic tube drum	62
Investigating how sound travels through air	63
Changing vibrations	64
Soundproofing investigation	65
Ear protection	67

Topic 6 The Earth and Space

The Earth's atmosphere	68
Water investigation	69
Investigate pollution	70
Model the water cycle	72
The water cycle	73
Reservoirs and dams	74
A dam or reservoir in my country	75
The Earth rotates (1)	76
Thinking about evidence: data about the Sun	77
The Earth rotates (2)	79
Investigation report: What would happen if the Earth was not tilted?	80
Why there are seasons on Earth	81
What do you know about satellites?	82

Appendix 1

Units for physical quantities	84

Topic **1** Our living world

Student's Book p 2
1.1 Planning a fair test

Planning a fair test

Plan a fair test to investigate what seeds need to start to grow (germinate).

1 Write the scientific question you will ask.

   _____

2 Which of the five main types of scientific enquiry will you use?

   _____

3 What dependent variable will you test?

   _____

4 How many independent variables will you test?

   _____

5 Make a list of what you will need to do a fair test.

   _____

6 Describe your control variables.

   _____

7 How will you measure what happens to the seeds? How often will you do this?

   _____

8 Is there anything you need to do to keep safe while you do this test?

   _____

9 Write the independent variable for each of the jars that you will use. Label each jar.

   Jar A: _____

   Jar B: _____

   Jar C: _____

   Jar D: _____

Topic 1 Our living world

The parts of plants (1)

Student's Book p 4
1.2 Not all plants produce flowers

1. Look at the picture of a fern leaf. Label and colour in the **spores**.

2. Describe where the spores are on the leaves.

_____

3. Look at the picture below. Describe what you can see.

_____

4. Draw some seeds to show where the seeds form in this plant.

Topic **1** Our living world

Student's Book p **4**
1.2 Not all plants produce flowers

The parts of plants (2)

1 Look at this picture of a flowering plant.

1 Label each part of the plant. Use the words in the box.

   leaf   flower   stem   root

2 Describe the job of each part of the plant.

Leaf _____

_____

Flower _____

_____

Stem _____

_____

Root _____

_____

3

Topic 1 Our living world

The parts of a flower

Student's Book p 6
1.3 The parts of a flower

This picture is a cross-section of a flower.

1. Use the words from your lists of the male and female parts of flowers to label the parts of the flower.
2. Colour in the picture.

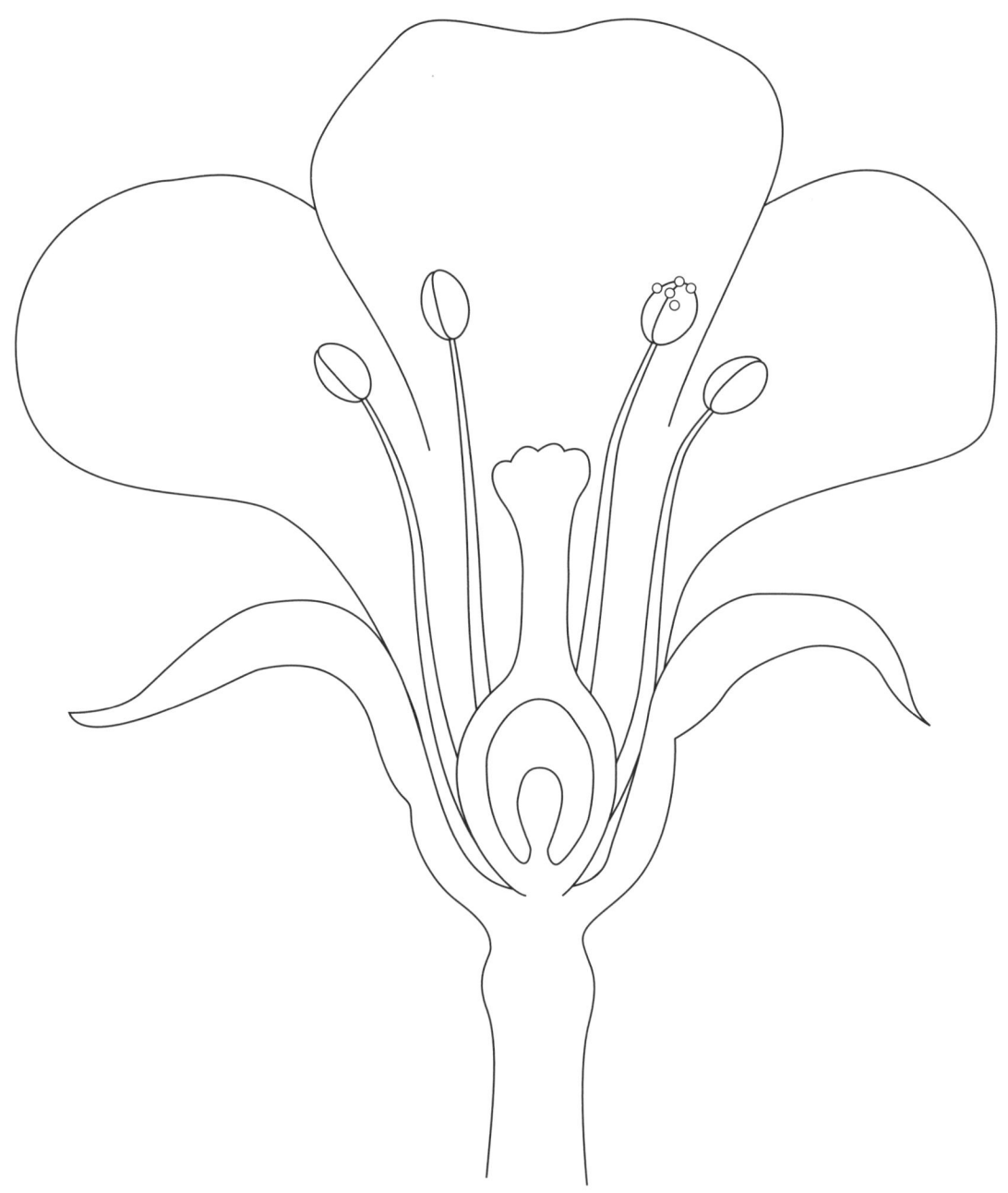

Topic **1** Our living world

Student's Book p 6
1.3 The parts of a flower

Drawing a cross-section of a flower

1 Write down the name of the flower that you investigated and describe what it looks like.

_____

_____

_____

2 Draw a picture to show what the flower looked like when it was cut open. Label all the parts.

Topic 1 Our living world

Student's Book p 8
1.4 From flowers to seeds

What happens during fertilisation?

1 Put these sentences about fertilisation into the correct order. The first one has been done for you.

Pollen goes into the ovary. ☐

Anthers make pollen. 1

Tube grows down the style. ☐

Pollen joins with the ovum. ☐

Seeds grow inside the ovary. ☐

Pollen lands on the stigma. ☐

2 Draw your own sketch to show how fertilisation takes place.

Topic **1** Our living world

Student's Book p **8**
1.4 From flowers to seeds

Investigating seeds

1. Collect at least six different types of seeds from your local environment.
2. Use this table to compare the seeds you have collected.

Picture of seed	Name of plant	Description of seed: • colour • size • hard or soft • type of protection

Student's Book p 10
1.5 Insects and flowers

Ways that flowers attract insects (1)

Examine four different flowers carefully.
Record your observations in the senses charts.
Use words such as 'sweet', 'soft', 'bright', 'pollen', 'petal' and 'silky'.

Name of plant: _____

Sense	Observation	Picture of flower
See		
Smell		
Feel		Size: _____ cm
Why do you think the flower is this shape and colour?		

Name of plant: _____

Sense	Observation	Picture of flower
See		
Smell		
Feel		Size: _____ cm
Why do you think the flower is this shape and colour?		

Topic 1 Our living world

Student's Book p 10
1.5 Insects and flowers

Ways that flowers attract insects (2)

Name of plant: _____

Sense	Observation	Picture of flower
See		
Smell		
Feel		Size: _____ cm
Why do you think the flower is this shape and colour?		

Name of plant: _____

Sense	Observation	Picture of flower
See		
Smell		
Feel		Size: _____ cm
Why do you think the flower is this shape and colour?		

Topic 1 Our living world

Insects and flowers

Student's Book p 10
1.5 Insects and flowers

1 What attracts insects to flowers? Name three things.

   _____

   _____

   _____

2 Draw sketches to show how pollination takes place.

3 Use your drawings to explain how flowers are pollinated by insects.

   _____

   _____

   _____

   _____

   _____

Topic **1** Our living world

Student's Book p 12
1.6 Seeds get around

Ways that seeds can be dispersed and spread around

1 Examine six different types of seeds.
2 Look at each seed carefully with a hand lens. Then discuss which way you think the seed could be dispersed – by the wind, by water, by animals, or by exploding.
3 Draw a picture of each seed and record your observations in the table below.

Picture	How this seed can be dispersed

Topic 1 Our living world

Student's Book p 12
1.6 Seeds get around

Animals that disperse seeds

Record your observations of animals in the table below.

1. Which animals did you observe? Remember that birds are also animals.
2. Record what each animal ate, and where and when it ate. Did the animal move around a lot?
3. What can you predict about the role of each animal in seed dispersal?

Picture and name of animal	Observations	Predictions about seed dispersal

4. Do you think you have collected enough data to draw a conclusion? Discuss this with your group.

Topic 1 Our living world

Student's Book p 12
1.6 Seeds get around

Making a model of a spinning seed

Some seeds are dispersed by the wind.
Some seeds have little wings like a helicopter.
These help the seed to spin away in the wind.

You can make a model of a spinning seed like this.

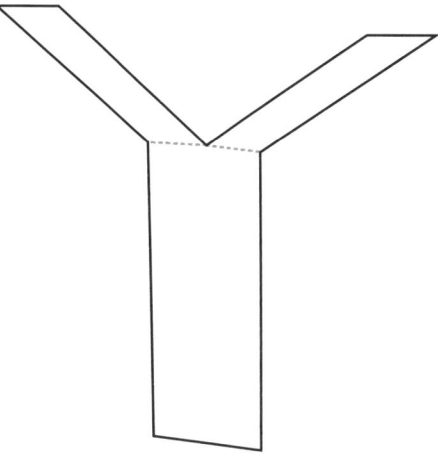

1. Copy this template on to a sheet of paper.

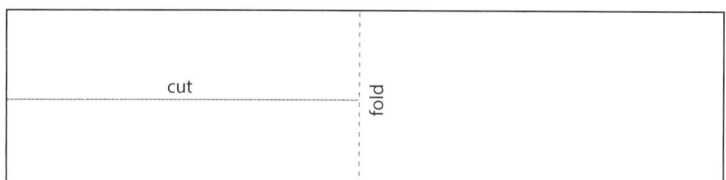

2. Cut it out and fold it to look like the picture.
3. Drop the spinner and watch what happens.
4. Then try dropping the spinner in a windy place. Watch what happens.
5. Add paperclips to the bottom of the spinner to make it heavier. Does this make a difference to how far or how fast the spinner moves?
6. You can also make the spinner bigger or smaller, or use a different type of paper. Does this make a difference?
7. Record your observations here.

_____

_____

_____

_____

_____

Topic 1 Our living world

Student's Book p 12
1.6 Seeds get around

What we found out about seeds with wings

Use your model of a spinning seed to answer these questions.

1 Did your spinner work well? _____

2 What did you notice about all of the spinners you made? _____
   _____

3 Did they spin clockwise or anticlockwise? _____

4 What can you do to change the spin? _____
   _____
   _____

5 Did the weight of the seed (the number of paperclips on the stem) change the spin?
   _____
   _____

6 What could you do to improve your spinner? _____
   _____

7 What would happen to the spinner if it was very windy? _____
   _____
   _____

Seed dispersal key

Follow the branches of the key to identify the plants.
Write the name of the plant in the correct box.

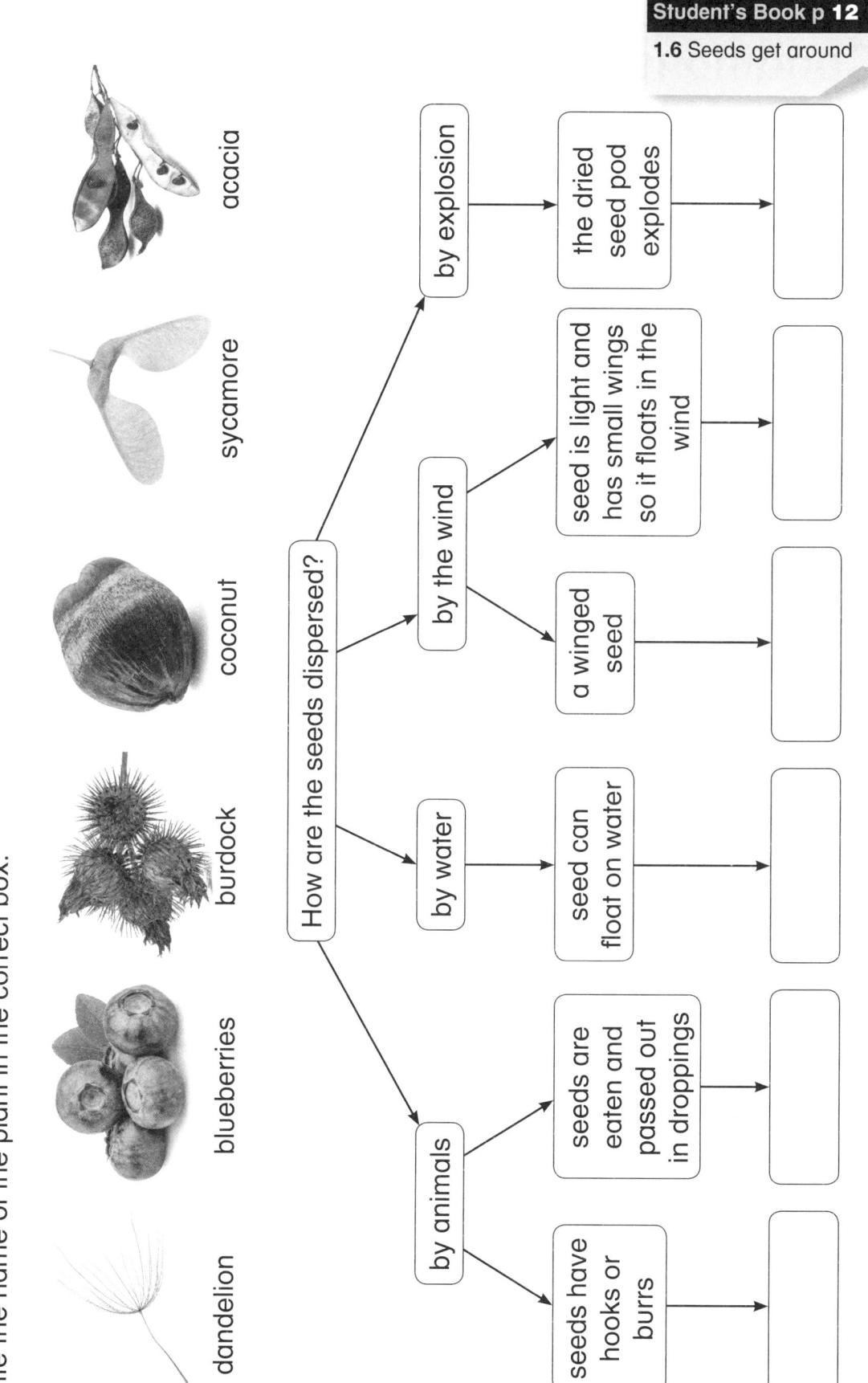

Topic 1 Our living world

Student's Book p 14
1.7 Germination

Conditions for germination

1 Record the independent variable in each jar in the table below.
2 What do you predict will happen to the bean seeds in each jar?
3 Observe what happens and record your observations in the table.

Jar	In warm or cool place?	Water or no water?	My prediction	Did the seed germinate?
A				
B				
C				
D				

Now answer these questions.

4 Did the bean seeds that you watered germinate?

_____

5 Did the bean seeds in the warm place germinate?

_____

6 Which bean seeds did not germinate?

_____

7 What can you conclude about the conditions that beans need to germinate?

_____

8 How accurate were your predictions?

_____

Topic **1** Our living world

Student's Book p 14
1.7 Germination

Seeds growing into plants

Each time you observe the plants:

- write the date
- draw a picture to show how the bean seeds are growing
- measure the height of the young plants
- write a few comments about what has happened.

Date: _____		Date: _____	
Height: _____		Height: _____	
Comments: _____		Comments: _____	

Date: _____		Date: _____	
Height: _____		Height: _____	
Comments: _____		Comments: _____	

Date: _____		Date: _____	
Height: _____		Height: _____	
Comments: _____		Comments: _____	

Stages in the life of a plant

Topic 1 Our living world

Student's Book p 16
1.8 The life cycle of a flowering plant

These pictures show stages in the life of a flowering plant.

1 Match the pictures to the phrases in the box below.

2 Write the correct phrase under each picture.

3 Number the pictures to show the order in which the stages occur.

| seedling grows   fruits and seeds are produced   shoot starts to grow |
| flower dies   flower forms   seeds are dispersed   seed germinates |
| seeds land on the ground |

Topic **1** Our living world

Student's Book p **16**
1.8 The life cycle of a flowering plant

The life cycle of a sunflower

Look at the stages in the life cycle of a sunflower in your Student's Book.

1 Make your own cycle diagram to show these stages.
Draw a picture of each stage and write a short description for it.

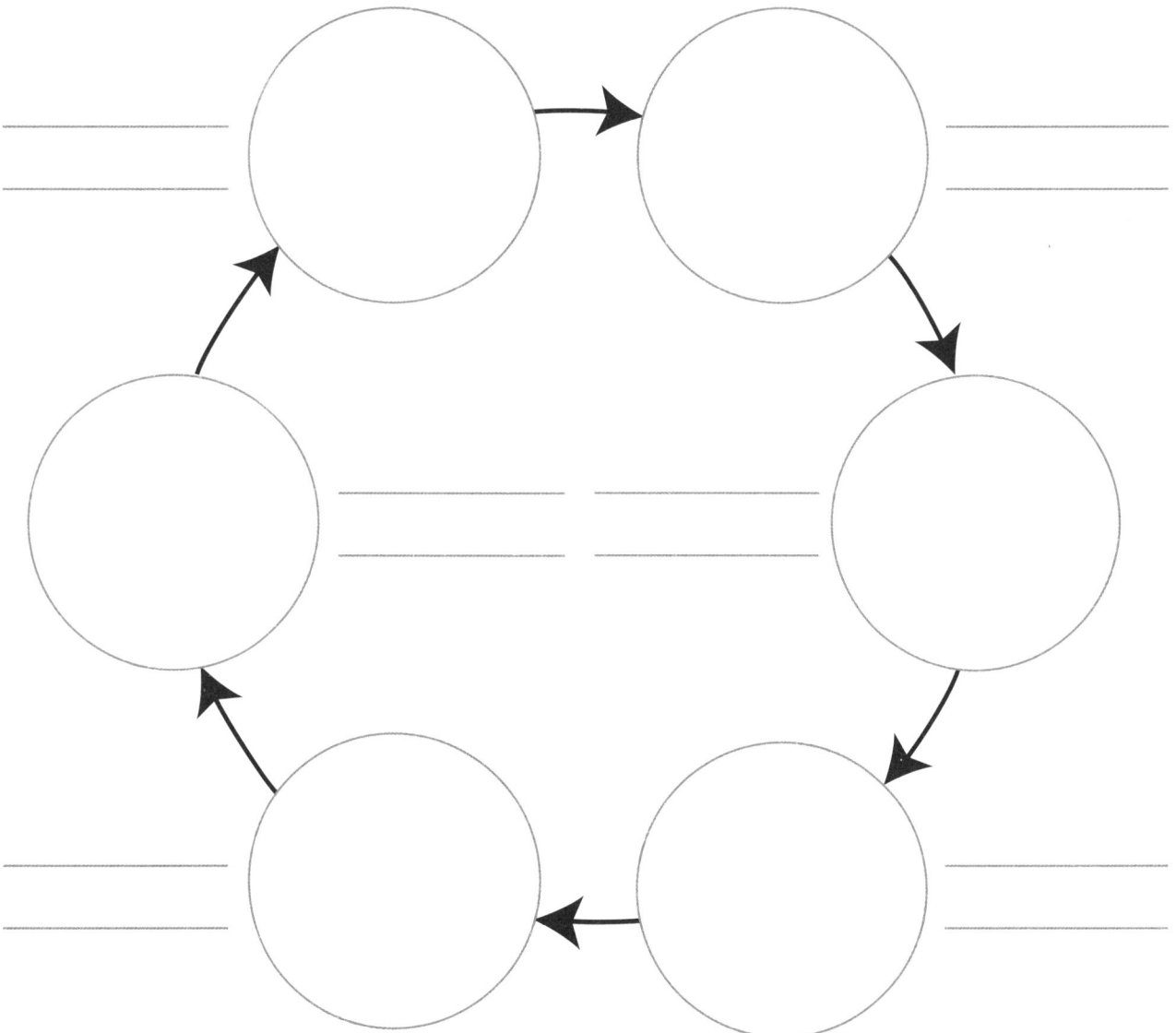

2 Describe ways in which you think sunflower seeds can be dispersed.

_____

_____

19

Topic 1 Our living world

Student's Book p 18
1.9 Plants and animals adapt

A plant that is adapted to its environment

1 Choose a plant. Draw a picture of the plant or find a picture and paste it here.

2 Label the main parts of the plant.

3 Describe the habitat of the plant.

4 Describe how the plant is adapted to its habitat.

Topic 1 Our living world

Student's Book p 18
1.9 Plants and animals adapt

An animal in a cold climate

Polar bear fur

You are going to use a container full of warm material to represent a polar bear. You will try at least three different materials as 'coats', to see which will keep the bear warmest for the longest time. You need to plan your investigation carefully, thinking about how to make it a fair test. What evidence will you collect and how often you will make your measurements?

1 Identify any control, independent and dependent variables.

_____

_____

2 What will I measure?

_____

3 How often will I take my measurements?

_____

4 What will I do to make the investigation a fair test?

_____

_____

5 I predict that:

_____

_____

_____

continued

Topic 1 Our living world

6 My results

Type of coat:	
Time (minutes)	Temperature (°C)

Type of coat:	
Time (minutes)	Temperature (°C)

Type of coat:	
Time (minutes)	Temperature (°C)

7 Do I need to repeat any of my measurements? Give a reason.

_____

8 My results tell me that _____

9 My prediction was _____

_____

continued

My procedure was:

1

2

3

4

5

6

Topic 1 Our living world

Predators and prey

Student's Book p 22
1.11 Predators and prey adapt

Look at this picture.

1. Which animal is the predator? _____
2. Which animal is the prey? _____
3. How is the predator adapted to catch its prey?

   _____

4. Complete these sentences. Use words from the box to help you.

   | run   prey   camouflage   adapted   predators   claws   spikes   smell |

   Animals that kill and eat other animals are called _____. The animals that they kill and eat are called _____.

   Predators are _____ to help them to catch their prey. For example, some predators have strong _____ and other predators can _____ very fast.

   Prey are also adapted so that they can avoid their predators. Some animals _____ themselves so that predators cannot see them easily. Other animals have _____ on their bodies or give off a _____ or poison so that predators do not attack them.

24

Topic 1 Our living world

Student's Book p 22
1.11 Predators and prey adapt

Predator and prey relationships

Coral is a colony of small animals that feeds on plankton. The crown-of-thorns starfish is a predator that feeds on coral.

Look at these four food chains from a coral reef.

plankton → coral → crown-of-thorns starfish → triton shell

plankton → coral → crown-of-thorns starfish → trigger fish

plankton → coral → crown-of-thorns starfish → Napoleon wrasse

plankton → coral → crown-of-thorns starfish → puffer fish

1. Name three predators of crown-of-thorns starfish.

2. The local community removes triton shells and sells them. Fishermen also catch trigger fish and Napoleon wrasse. In what way do you think this affects the starfish numbers? Why?

3. Complete this flow diagram to show what happens on the reef if the number of starfish increases:

| predator numbers increase | → | they eat more | → | number of prey |

4. The government makes part of the reef a conservation area and bans shell collecting and fishing in that area. As a result, the number of trigger fish increases quite quickly. Draw a flow chart like the one above to show the impact this could have on the numbers of prey.

5. In some areas, starfish numbers have increased dramatically and they are destroying corals. Many conservation organisations pay divers to find and remove them. Why do you think it is important to remove predator starfish from coral reefs?

Topic 2 Healthy eating

The organs of the human digestive system

Student's Book p 28
2.2 The human digestive system

Label the parts of the human digestive system. Use words from the box.

| anus   oesophagus   small intestine   large intestine   stomach   mouth |

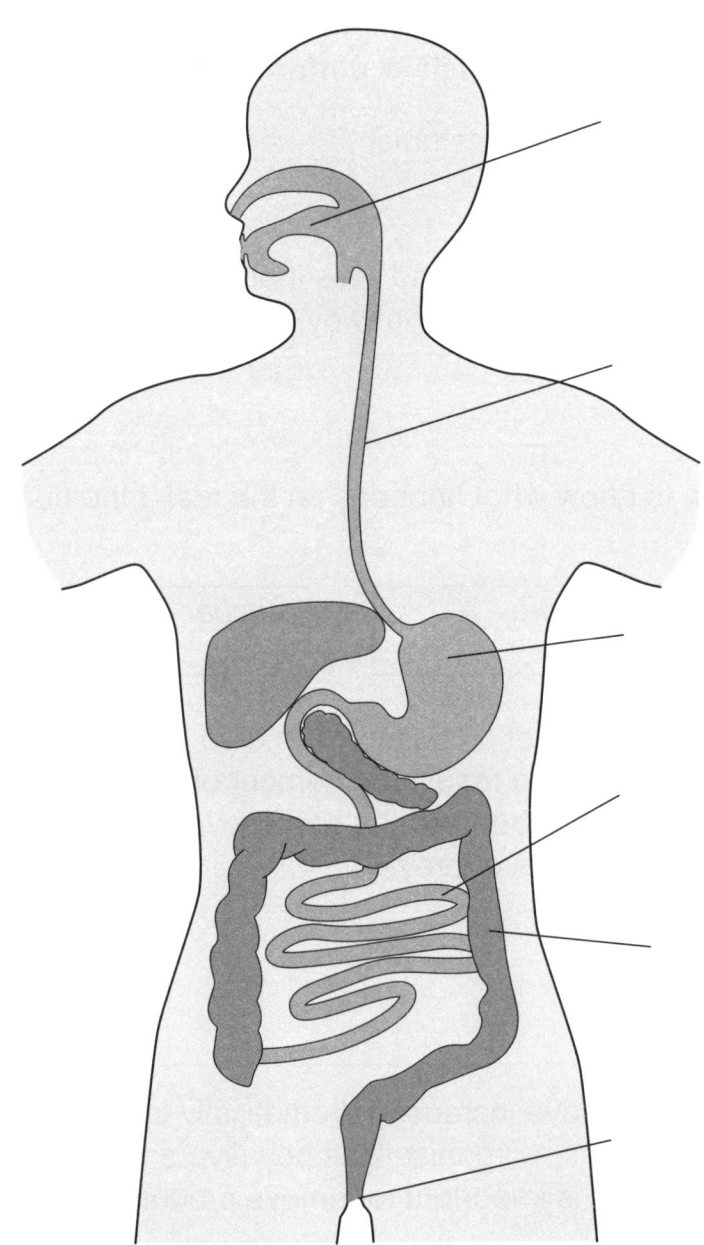

Topic **2** Healthy eating

Student's Book p 28
2.2 The human digestive system

How food moves through the digestive system

Put this information in the correct sequence to show how food moves through the human digestive system. Label each organ in the system. Add arrows to show the movement.

Food is absorbed into the blood.

Waste is formed.

Food is chewed and swallowed.

Food moves down into the stomach.

Food is crushed and made into liquid.

Waste passes out of the body.

Topic 2 Healthy eating

Food pyramid diagram

Student's Book p 30
2.3 A balanced diet

1. A food pyramid shows us how much of each food group we should eat.
   - Make a list in your notebook of the food you ate yesterday.
   - Write what food group each belongs to.
   - Write or draw the foods into the correct place on the food pyramid diagram.

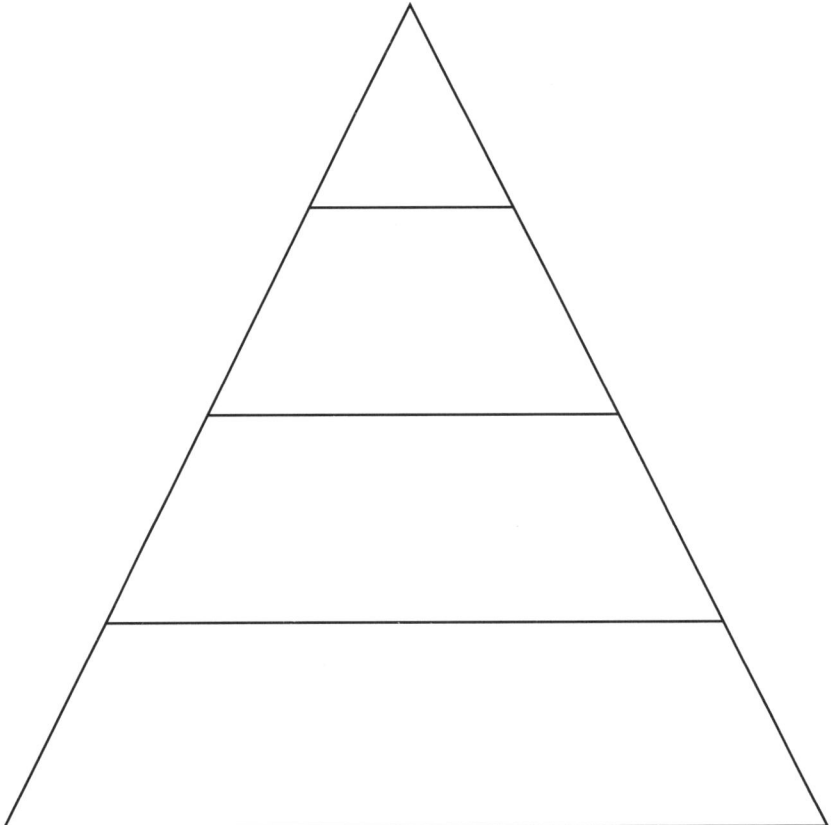

2. What food groups did you eat the most of?

_____

3. Which food groups do you think you should eat more of?

_____

4. How does using a food pyramid diagram help you to eat a balanced diet?

_____

A balanced meal

Topic 2 Healthy eating

Student's Book p **30**
2.3 A balanced diet

1 Plan a balanced meal. Draw the meal on the plate.

2 Which food groups have you included on the plate?

3 Which foods take up the most space on the plate? Why?

Topic 3 Materials

Student's Book p 38
3.2 The particle model

Diagrams show what matter looks like

1

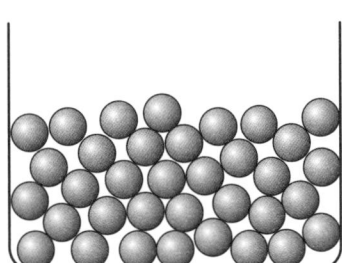

This diagram shows the particles in a _____.

The particles are _____.

2

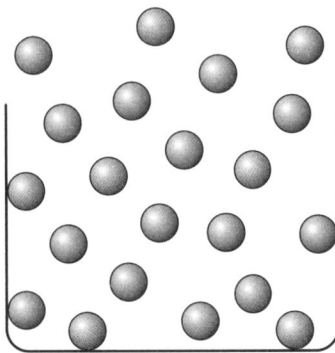

This diagram shows the particles in a _____.

The particles are _____.

3 Draw a diagram to show the particles in a solid.

4 Describe how the particles in a solid are arranged.

_____.

30

Topic 3 Materials

Student's Book p 38
3.2 The particle model

How matter behaves and changes state

The particle model explains how matter behaves and changes state.

1 Complete these sentences.

| move   spaces   attracted   particle   changes   particles   temperature |

The _____ model explains how matter _____

state and how it behaves. All matter is made up of _____.

Particles _____ all the time. There are _____

between the particles. The particles are _____ to each other.

_____ can cause matter to change state.

2 Look at the diagram. Do you think the particles in this matter will be able to move easily or not? Give a reason for your answer.

3 The particles in a gas move around freely. Explain, using a diagram, why you think the particles are able to do this.

Topic 3 Materials

Student's Book p 42
3.4 Gases, solids and liquids

States of matter diagram

1. Make a concept map of the states of materials and their properties. Use information from your Student's Book, and from your own discussions and investigations.

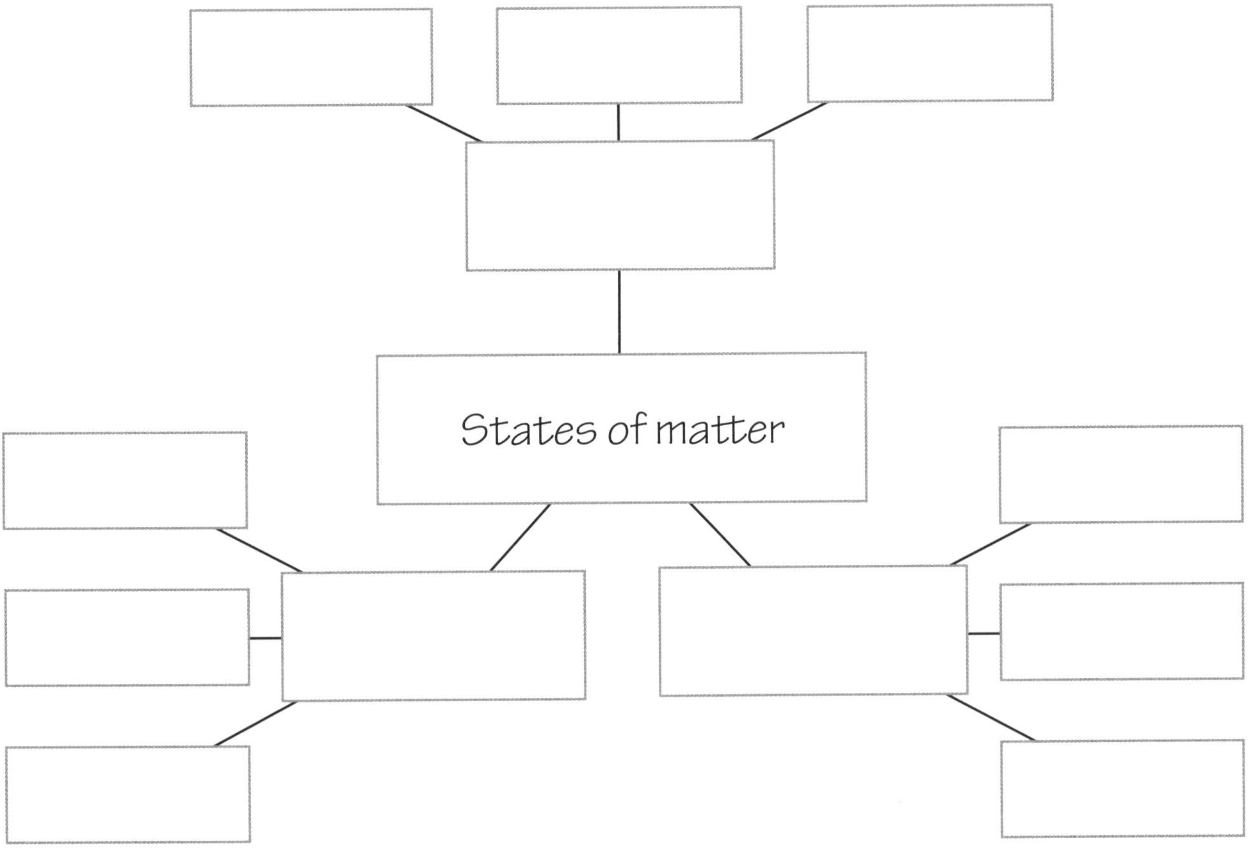

[States of matter]

2. Write down the state of water in each of these photographs.

Topic 3 Materials

Student's Book p 42
3.4 Gases, solids and liquids

A solid changes to a liquid

1 Draw a diagram to show how particles are arranged in a solid block of ice.

2 Now show what the particles look like when the ice changes to a liquid.

3 What can cause a solid to change into a liquid?

_____.

4 Describe what happens to the particles when the solid changes to a liquid.

_____

_____.

Topic 3 Materials

Puddle investigation

Student's Book p 44
3.5 Evaporation

Plan your own puddle investigation to see how quickly two different puddles take to evaporate.

1 What scientific question will you ask?

2 Which of the five main types of scientific enquiry will you use in your investigation?

3 Write your plan.

4 Describe the independent variables.

5 What do you predict will happen?

6 Record your results.

7 Was your prediction correct?

8 What could you do to improve your investigation?

9 Was your investigation a fair test? If not, say why.

Topic **3** Materials

Student's Book p **44**
3.5 Evaporation

Drying clothes

Plan and record an investigation to find out a way to dry clothes quickly.

Your teacher will give you:
- three T-shirts (or pieces of fabric), A, B and C
- water
- weighing scales
- washing line
- pegs

1. Use the equipment your teacher has given you to plan your investigation.
2. What scientific question will you ask?

3. Which of the five main types of scientific enquiry will you use in your investigation?

4. Describe the dependent variable.

5. Describe the independent variables.

6. What is the control variable? How will you make this a fair test?

7. Write your plan.

8. What do you predict will happen?

continued 35

Topic 3 Materials

9 Record your results.

T-shirt	Where did you hang it?	Mass before drying (g)	Mass after ___ hours (g)	Difference in mass (g)
A				
B				
C				

10 Draw a bar chart to show the results.

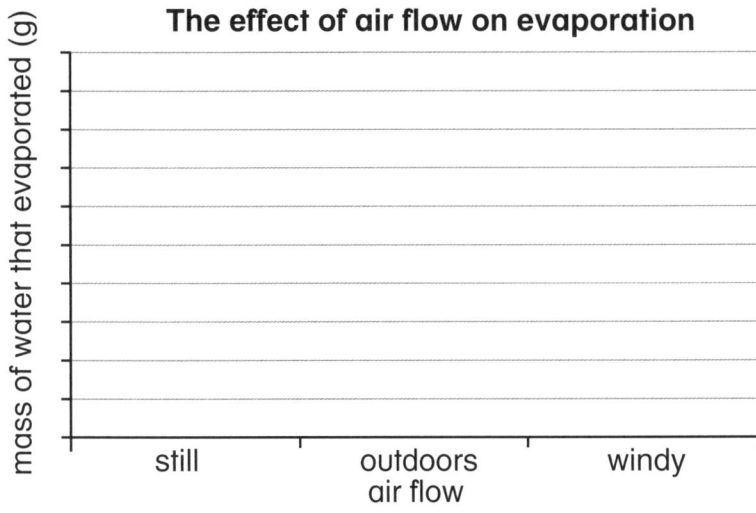

11 What do the results tell you? How accurate were your predictions?

_____

_____

12 Was this a fair test?

_____

_____

13 Do you think air flow was the only factor that affected the rate of evaporation?

_____

_____

Topic 3 Materials

Student's Book p 46
3.6 Condensation

Investigating condensation

When water vapour in the air cools it condenses. Follow the instructions below to observe this change of state.

You will need:
- an empty food can
- ice cubes
- a dry cloth

1 Wipe the can with a dry cloth.

2 Fill the can with ice.

3 Wipe the can again.

4 Leave the can in a warm place and observe what happens

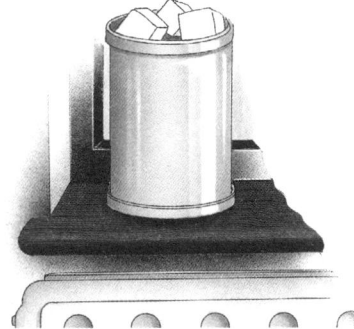

Where does the water on the outside of the can come from?

_____

_____

Use the particle model to explain the change of state. Explain what role you think temperature has in this investigation.

_____

_____

Topic 3 Materials

Collecting water

Student's Book p 46
3.6 Condensation

Look at this picture of a scientist using a dew collector in the Atacama Desert. ▶

A dew collector is made from a large net connected to a funnel and a container. Deserts are hot during the daytime but are cold at night. At night, the water vapour in the warm air cools and changes state from a gas to a liquid. Water forms as condensation on the net of the dew collector. The water runs downwards into the funnel and by morning the container is full of fresh drinking water.

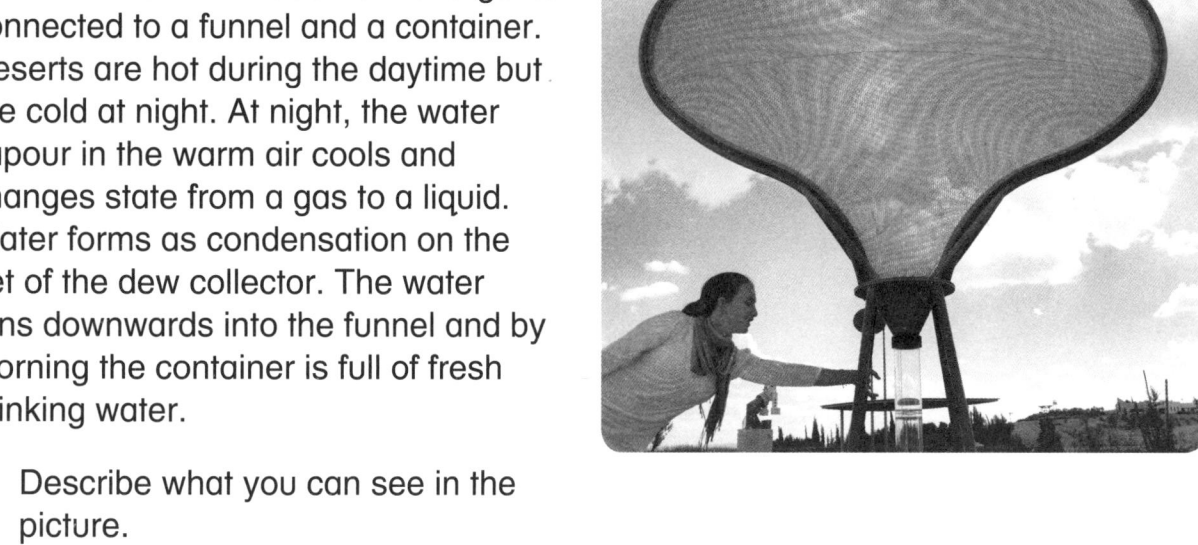

1 Describe what you can see in the picture.

_____

_____

_____

_____

2 Where does the water in the container come from?

_____

_____

3 Use the particle model to explain the changes of state that are happening.

_____

_____

4 Where could this sort of equipment be useful?

_____

_____

Topic 3 Materials

Student's Book p 48
3.7 Water boils and freezes

Heating ice

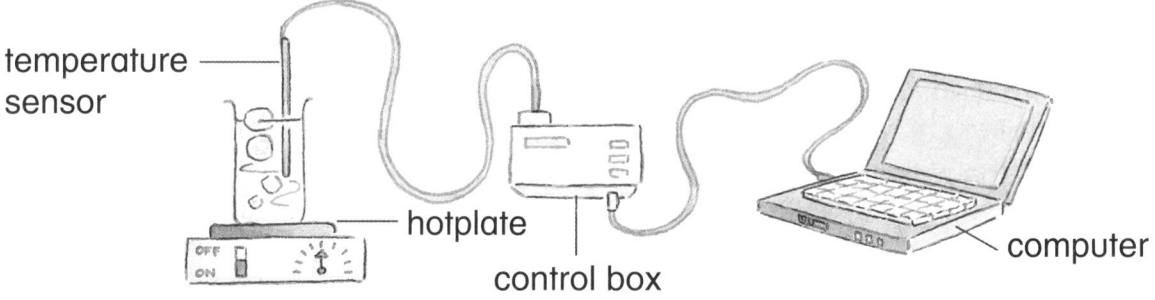

1. Look at the picture. ▲ What scientific question is being investigated? Describe what was used in the investigation.

2. Look at the graph. How many minutes did it take for the ice to reach melting point?

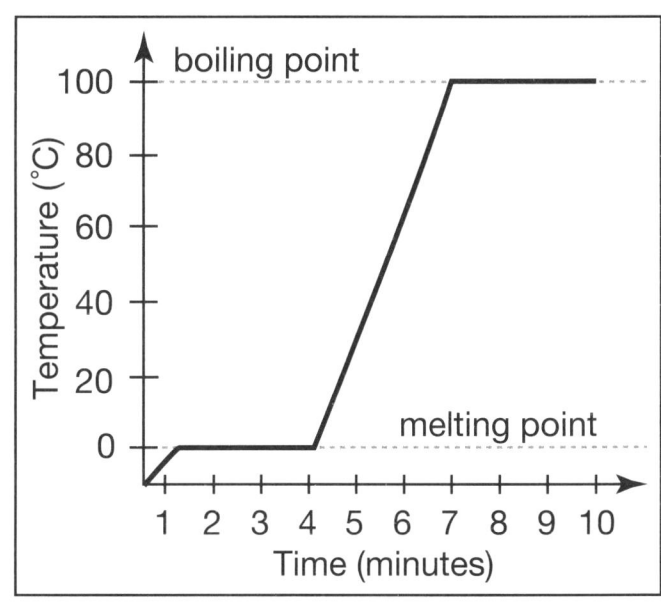

3. At what temperature did the ice reach melting point?

4. For how long did the temperature of the water stay at melting point?

5. How many minutes did it take for the water to reach boiling point?

6. At what temperature did the water reach boiling point?

7. Did the temperature change after the water reached boiling point?

8. Write a heading for the graph.

39

Topic 3 Materials

What happens when you heat ice?

Student's Book p 48
3.7 Water boils and freezes

Plan your investigation into the way that the temperature of ice changes when it is heated, and to record the results. Before you begin, discuss with your partner how to minimise any risks during this practical.

You will need:
- a pot of ice
- a thermometer
- a heater and support
- a stopclock or watch with a timer

NOTE: Always be careful when you heat something.

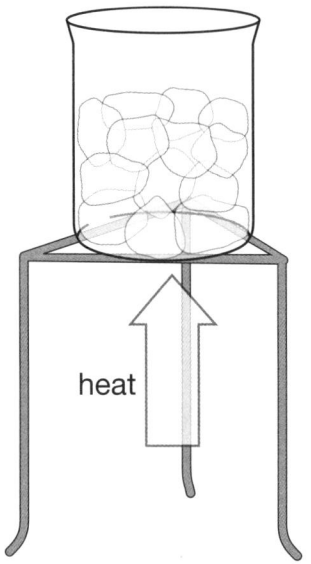
heat

1. Measure and record the temperature of the ice before you start.
2. Put the pot of ice on the heater. Be careful!
3. Start the timer. Heat the ice SLOWLY, using the thermometer to gently stir it all the time.
4. Record the temperature every 30 seconds.
5. Stop heating when all the ice has melted.
6. Complete this sentence:

   The ice melted at a temperature of _____.

Time (minutes)	Temperature (°C)
0 seconds (start)	
30 seconds	
1 minute	
1 minute 30 seconds	
2 minutes	
2 minutes 30 seconds	
3 minutes	
3 minutes 30 seconds	
4 minutes	
4 minutes 30 seconds	
5 minutes	

continued

7 Draw a graph to show the results. Record the temperature on one axis and the time on the other axis.

Temperature (°C)

Time (minutes)

Does sugar dissolve?

1 Write the scientific question you will investigate.

2 Which of the five main types of scientific enquiry will you use?
Draw and label a picture to show the equipment you will use.

3 What will you do to make this a fair test? Describe the dependent and independent variables.

4 What do you predict will happen?

5 Explain the steps in your investigation.

6 What did you find out?

Topic 3 — Materials

Student's Book p 50
3.8 Mixing and dissolving matter

Temperature and dissolving

Set up an investigation to find out if temperature affects the rate at which sugar dissolves.

1 What is your prediction?

2 Explain briefly how you will investigate this. Draw and label a picture to show the equipment you will use.

3 Explain what measurements you will take and how you will record the measurements.

4 How will you show your results? Show the result here or record them on a sheet of paper.

Making crystals

Student's Book p 52
3.9 Separating solvents and solutes

You are going to investigate a way to make crystals using the process of evaporation.

Use this sheet to plan your investigation and to record the results.

NOTE: Always be careful when you heat something.

1 Look in your Student's Book and make a list of the equipment you will need for this investigation.

_____

_____

2 Make the saturated solution first. Write down notes about what you will do.

_____

_____

3 Make the crystals. Then draw a picture of the crystals that you made. Label your picture.

4 Describe the size and shape of the crystals.

_____

_____

5 Did you do anything to change the shape and size of the crystals? For example, did you heat the saturated solution while the crystals were forming? If so, what effect did this have?

_____

_____

Topic **4** Forces

Student's Book p 56
4.1 Recording observations

Use a table to record observations

1 Describe the four objects that you will test in water.

Object 1 _____

Object 2 _____

Object 3 _____

Object 4 _____

2 Now add headings to this table so that you can record your observations. Don't forget to give the table a heading and to record the date of your investigation

Investigation: _____

3 How do you think you could improve this investigation?

_____

_____

 Topic 4 Forces

Student's Book p 56
4.1 Recording observations

Using a line graph

Use a line graph to show the results of an investigation.

1 Draw a line graph to record the data that your teacher gives you. Do not forget to label each axis and to give the graph a title.

2 How does this graph help you to interpret the results of the investigation?

_____

3 What can you conclude from this investigation?

_____

Topic 4 Forces

Student's Book p 58
4.2 Different forces

What do you know about forces?

1 Are the following sentences true or false? Correct the sentences that are false.

A force is a push or a pull on an object. True ☐ False ☐

_____

Forces occur when things crash into each other. True ☐ False ☐

_____

Forces can work in different directions. True ☐ False ☐

_____

Forces are visible. True ☐ False ☐

_____

2 Add arrows to this diagram to show the direction in which forces are exerted so that this box stays on the table.

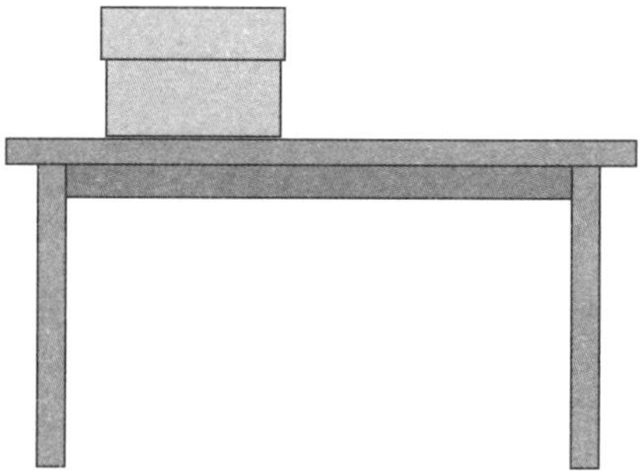

3 Label the forces in the diagram above. Choose words from the box.

| gravitational pull   normal pull |

47

Topic 4 Forces

Student's Book p 60
4.3 Friction

Investigate friction on different surfaces

Your teacher will give you:
- A wooden board
- A smooth plastic tray

You need to choose:
- Four items that won't break easily (e.g. an eraser, a matchbox, a coin, a plastic cup, a stone)

1 What scientific question do you want to ask?

2 Which of the five main types of scientific enquiry will you use in your investigation?

3 How will you set up the equipment you have to answer your scientific question?

4 Predict which objects will slide down each surface the fastest and which will slide the slowest if you tilt the surface at an angle.

My predictions	Wooden surface	Plastic surface
Will slide fastest		
Will slide slowest		

5 Were your predictions correct?

6 What could you do to adapt this experiment to make it a fair test?

7 Explain why some objects slide more easily than others do.

Topic **4** Forces

Student's Book p **60**
4.3 Friction

Investigating friction

Your teacher will give you:
- a toy car
- some books
- a timer
- a ramp and some carpet, sandpaper and corrugated cardboard

1 Plan and set up an investigation to see what affects friction.

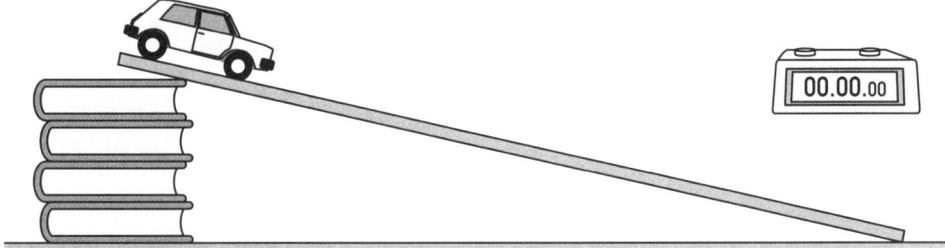

2 What scientific question will you ask?

_____

3 Which of the five main types of scientific enquiry will you use?

_____

_____

4 What is the control variable in your investigation?

_____

5 What is the dependent variable?

_____

6 What is the independent variable?

_____

7 How will you make this a fair test?

_____

_____

_____

continued ➡ 49

Topic **4** Forces

8 Predict which surface you think will have the fastest time. Why do you think this?

_____

_____

9 Fill in the table with your results. Test each surface three times.

Surface	None (control)				
Time 1					
Time 2					
Time 3					

10 How accurate are your results? Are there any results that do not fit the pattern? Do you need to do any of the tests again?

_____

_____

_____

11 What did you find out? How does the surface affect the time it takes for the car to roll down the ramp? Explain your results.

_____

_____

_____

Topic 4 Forces

Student's Book p 62
4.4 Air resistance

Parachute investigation

> **You will need:**
> A small parachute made from a 50 cm square, thin polythene sheet (e.g. bin liner) or thin cloth (e.g. silk or nylon). The strings can be made from thick thread, tied to each corner. The mass could be modelling clay or a toy figure of mass about 25 g.

Investigation 1

- Fold the parachute and fasten it to the weight with an elastic band.
- Time how long the parachute takes to fall a distance of 5–10 metres without opening.
- Improve the accuracy of your timing by counting down before letting go of the parachute, e.g. 'three – two – one – drop'.

Investigation 2

- Drop the parachute from the same height, but this time allow it to open.
- Time how long it takes to fall.

Design and make a results table in the space below. Record your results.

Topic 4 Forces

Student's Book p 62
4.4 Air resistance

Investigate air resistance on cars

The car in picture **A** is the Bloodhound SSC. It is a supersonic car, which means that it can travel faster than sound. This is much faster than a normal car!

A

B

1. Discuss how the car in picture **A** looks different from the car in picture **B**.

2. Which car do you think has the most air resistance acting on it when it moves?

   _____

3. Use the shape of the cars to explain why you think this.

   _____

   _____

4. Work in a team to plan and carry out an investigation to find out how air resistance slows cars down. Use the picture on the left for ideas.

continued

5 Record the results of your investigation here.

_____

_____

_____

_____

_____

6 How did you change the amount of air resistance acting on the car?

_____

_____

7 Use the results from your investigation to explain why the shape of Bloodhound SSC helps it to go so fast.

_____

_____

_____

_____

8 Is there anything you could change to improve the results of your investigation?

_____

_____

Does shape affect water resistance?

Student's Book p 64
4.5 Water resistance

Work in groups and design an object that sinks quickly to the bottom of a bottle of water. You can use clay to make objects. Mould the clay into the shapes that you think will sink the most quickly.

1 What shapes do you predict will sink quickly?

_____

2 How can you make this a fair test?

_____

3 Conduct your investigation. Draw a diagram of the shape that sank the most quickly here. Draw arrows to show the direction of the forces on the object in the water. Label the upthrust and the weight.

4 Write your conclusion here. Explain why this shape sank the most quickly.

_____

_____

Topic 4 Forces

Student's Book p 64
4.5 Water resistance

Design a toy which travels through water

A toy company wants to make a fun new toy, which moves quickly underwater. Because you have learned all about water resistance, they have asked you to design it!

Draw your design in the box below. Label the important parts to explain how it will work.

Things to consider:

- How will your toy be powered: will it be wind-up or battery?
- What will it be made of?
- What will it be: an animal, a vehicle or something else?
- How will you make sure it swims quickly through the water?
- Do you want to be able to change the direction of the toy as it swims through the water?

Topic 4 Forces

Student's Book p 66
4.6 Magnets and magnetic materials

Magnetic or non-magnetic?

Design an experiment to find out which metals are attracted to a magnet. You can use things like drink cans, jewellery and tools in your experiment.

1 What do you want to find out?

___

2 What will you need?

___

3 What will you do?

___

4 Use this table to organise your results.

Item tested	Metal it is made from	Magnetic or non-magnetic

5 Make a list of metals that are magnetic. Use your experiment results but also carry out research to find out about other metals.

___
___
___

Topic 4 Forces

Student's Book p 66
4.6 Magnets and magnetic materials

Are the poles of a magnet equally strong?

Design a fair test to find out whether the poles of a magnet are equally strong.

1 What are you trying to find out?

2 What apparatus will you use?

3 What will you do?

4 What will you do to make your test fair?

5 What will you measure?

6 What do you predict will happen?

Topic 4 Forces

Magnetic forces

Student's Book p 68
4.7 Investigate magnetic forces

Use two bar magnets to investigate the distance over which they attract (pull) and repel (push) one another.

You will need:
two bar magnets and a ruler.

1. Place the ruler on a flat surface.
2. Position one of the bar magnets at 0 cm as shown below. Hold the magnet down firmly so it does not move.

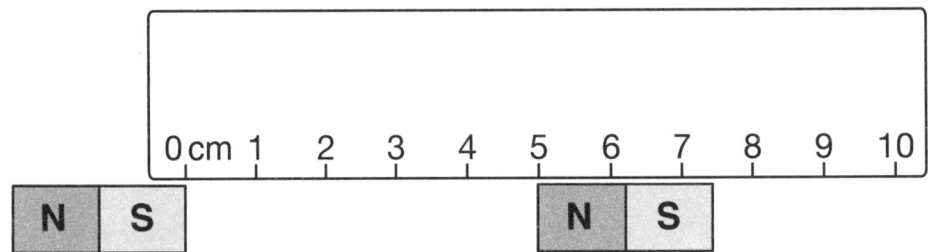

3. With opposite poles together, gently slide the second magnet towards the magnet which is being held down.
4. At what distance did the like poles begin to attract?

_____

5. What do you think would happen if you used larger magnets? Does the distance over which the forces act increase or decrease?

_____

6. Repeat the investigation but this time with the same poles of the magnets together.
7. At what distance was the second magnet when it stopped being repelled?

_____

8. What happens to the forces acting on the magnets? Do they become weaker or stronger?

_____

Topic **4** Forces

Student's Book p **68**

4.7 Investigate magnetic forces

Forces and distance

Show how forces act over distance between a magnet and a magnetic material.

You will need:
bar magnet, two paperclips and a ruler.

1 Place the ruler on a flat surface.

2 Position the bar magnet at 0 cm as shown below. Hold the magnet down firmly so it does not move.

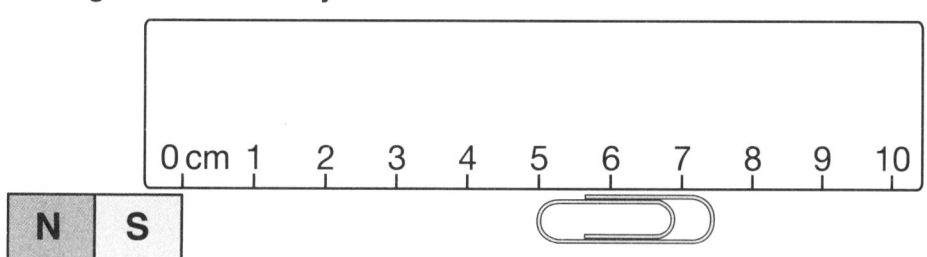

3 Start by holding the paperclip at 30 cm. Slowly move the paperclip towards the magnet.

4 Record the distance at which the paperclip was attracted (pulled) towards the magnet.

_____

5 What do you think would happen if you used larger magnet?

_____

6 Repeat the investigation but this time with two paperclips joined together.

7 Record the distance at which the paperclips were attracted (pulled) towards the magnet.

_____

8 Is the distance the same for one and two paperclips? Why is this?

_____

9 Do you think you would get the same result for 10 paperclips?

_____

59

Topic 5 Sound

The string telephone

Student's Book p 72
5.1 Vibrations

You are going to work with a partner to design and make a string telephone.

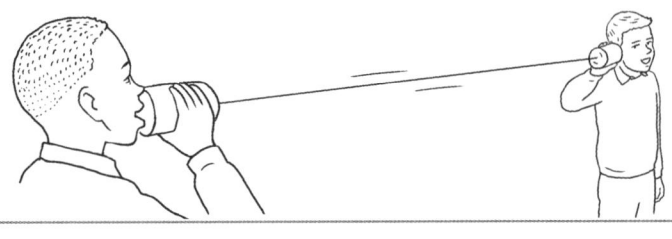

1 Draw your design here.

2 Label your diagram to show:
- the materials used for the cup and the string
- the length of string you plan to use
- the size of the cups
- how you will stop the string from pulling loose.

3 Predict whether the telephone will work better with the string loose or tight.

4 Build the telephone you designed and try it out.

5 Record how well your telephone worked.

continued

Topic **5** Sound

6 Explain how sound vibrations travel in the string telephone.

_____

_____

7 Explain what you could change to improve your investigation.

_____

8 Could you improve your design? Make two suggestions.

_____

_____

9 These are some questions that students asked about string telephones:

- How does the length of the string affect the sound?
- What is the longest piece of string we can use and have the phone still work?
- Do the cups have to be the same size?
- Is it possible to connect three people and have the phone still work?
- Do smaller or bigger cups work better?
- Would the sound be better if we used tins instead of paper cups?

a Choose one question that interests you.

Predict the answer.

_____

b Plan an investigation to see whether your prediction is correct. Carry it out and record your findings below.

What we did: What we learned:

_____   _____

_____   _____

Topic **5** Sound

Plastic tube drum

Student's Book p 72
5.1 Vibrations

Stretch some plastic, paper or cling film over a tube and hold it in place with an elastic band. Place a few rice grains on the plastic or paper skin.

1 Draw what happens to the rice grains when you make a quiet noise near the bottom of the tube.

2 Draw what happens to the rice grains when you make a loud noise near the bottom of the tube.

3 Write an explanation of the behaviour of the rice grains that you have drawn above.

_____

_____

_____

Topic 5 Sound

Student's Book p **72**
5.1 Vibrations

Investigating how sound travels through air

You are going to do an experiment to show that sound vibrations travel through air.

> **You will need:**
> - a plastic bottle
> - a plastic bag or plastic wrap
> - an elastic band
> - scissors
> - a small candle

Follow these steps:

Step 1: Carefully cut the bottom off the bottle.

Step 2: Stretch a piece of plastic film tightly over the open end of the bottle. Secure it with the elastic band.

Step 3: Light the candle. Hold the bottle so the neck is about 3 cm away from the candle flame. Be very careful with the flame!

Step 4: Hold the bottle very still and tap the plastic film with your fingertips.

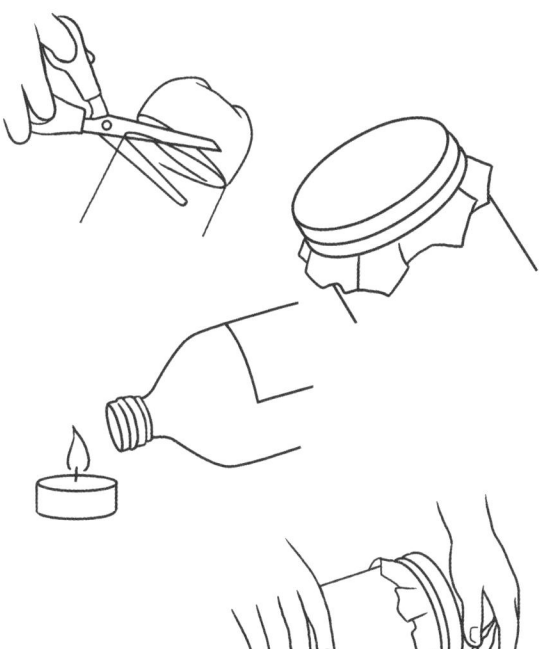

1 What happened to the candle flame?

_____

2 Why do you think this happened?

_____

3 Draw a labelled diagram in your exercise book to show what happened in this experiment. Use arrows to show the movement of air.

Topic 5 Sound

Student's Book p 74
5.2 Volume and pitch

Changing vibrations

1. Hold a plastic ruler over the end of your desk, as shown in the diagram.
   Keep one hand on the ruler on the desk.
   Lift up the loose end and let it go.

   a What happens to the loose end of the ruler?

   _____

   b When does the sound stop?

   _____

2. Repeat the experiment with the ruler a few times, changing the length of the section that hangs over the desk.

   a Which length of ruler produces the fastest vibrations?

   _____

   b Which length of ruler gives the highest sound?

   _____

   c What could you do to make the ruler vibrate slowly and produce a low sound?

   _____

   d What does this experiment show?

   _____

   _____

Topic 5 Sound

Student's Book p 74
5.2 Volume and pitch

Soundproofing investigation

1 What scientific question will you ask?

2 Which of the five main types of scientific enquiry will you use?

3 Identify any variables in your investigation.

4 What will you do to make your test fair?

5 Results table

Material	Description of sound level	Sound level in decibels (dB)			
		Try 1	Try 2	Try 3	Average
None (control)					

6 Which material was best at soundproofing?

continued

Topic 5 Sound

7 Draw a bar graph of your results.

Topic **5** Sound

Student's Book p **76**
5.3 Controlling sound

Ear protection

Make and test a set of ear protectors using cups filled with three different materials.

Measure how well each set of ear protectors works to reduce sound by asking someone to turn up the volume of the TV or radio one step at a time until you can hear it.

1 To make your test a fair test, you will need to stand with your back to the source of the sound and in the same position for each set of ear protectors. Use this table to record your findings.

	Materials tested	Volume at which the sound could be heard
Set A		
Set B		
Set C		

2 a Which set of ear protectors was the most effective at reducing sound?

b Why do you think this was the case?

c Explain what you could change to improve your investigation.

d Could you improve on your design? Say how.

Topic 6 The Earth and Space

Student's Book p 80
6.1 The Earth's atmosphere

The Earth's atmosphere

1 Complete these sentences.
The atmosphere around the Earth is made up of _____ of different gases. The layers of atmosphere nearest the Earth is made up of _____ and _____. These gases keep the Earth at the right _____ for us to be able to live. The atmosphere also protects us against _____ from the Sun.

Human activities on Earth change the atmosphere. For example, we burn _____, which is a fossil fuel. This causes more of the gas _____ _____ to go up into the atmosphere.
People also cut down trees, which change harmful carbon dioxide gas into _____, which is a gas that living things need.

2 Draw a pie chart to show how much oxygen, nitrogen and carbon dioxide are found in the Earth's atmosphere.

3 Explain how this pie chart could change if humans continue to burn fossil fuels and cut down trees without replacing them.

_____

_____

Topic **6** The Earth and Space

Student's Book p 82
6.2 Water everywhere

Water investigation

You are going to test a sample of water to find out if it is from the ocean (salty) or from a river (fresh water).

1 List the materials you will need.

_____

_____

2 Describe what you will do. What process will you use?

_____

_____

_____

_____

3 Test your sample of water.

4 Record your observations.

_____

_____

_____

_____

5 What is your conclusion?

_____

_____

_____

_____

Topic 6 The Earth and Space

Student's Book p 84
6.3 Pollution

Investigate pollution

You will need:
- card
- hole punch
- scissors
- string
- petroleum jelly
- hand lenses

Investigate the amount of air pollution in different places around your school by making particulate collection cards.

Setting up the equipment

For each place that you choose to investigate, make a collection card by following these steps:

a Use the hole punch to make a hole near the corner of the card.

b Thread some string through the hole.

c Label the card with your names and where you are going to place it.

d Attach the card by the string to the place you are investigating.

e Smear petroleum jelly evenly, all over the card.

f Once you have made your cards, work in groups to plan your fair test investigation and answer the questions on the next page.

1 Where will you place your cards?

   A _____   B _____   C _____   D _____

2 What will you measure at the end of the investigation?

   _____

   _____

3 What is the control variable in this investigation?

   _____

   _____

continued

Topic 6 The Earth and Space

4 Make a prediction about where you think you will find the most pollution. Explain why you think this.

_____

_____

5 You will need to retrieve your cards after about one week.

6 Carefully count the particulates on each card. Use a hand lens to help you do this. Record your results in the table below.

Place	Number of particulates
A	
B	
C	
D	

7 Which place around the school had the most particulates?

_____

8 Explain why you think this happened.

_____

_____

9 Was your prediction correct?

_____

10 Did any results surprise you?

_____

_____

11 Explain what you think the particulates were.

_____

_____

Topic 6 The Earth and Space

Student's Book p 86
6.4 The water cycle

Model the water cycle

You will need:
- a large glass or plastic bottle with a lid
- equal amounts of small pebbles, soil and sand
- a small green plant that will fit inside the jar
- a small container for water
- a sunny window ledge

1. Put the small pebbles into the bottle first. Then put the sand on top of the pebbles and the soil on top of the sand.
2. Put the small plant into the soil.
3. Put water in the small container and put it into the bottle, next to the plant.
4. Put the lid on the bottle. (Use sticky tape to close the bottle if you do not have a lid.)
5. Put the bottle in a sunny place for a few days and observe what happens. Record your observations here and explain what happened.

_____

_____

_____

_____

_____

_____

Topic 6 The Earth and Space

Student's Book p 86
6.4 The water cycle

The water cycle

1 Label this picture so that it shows the water cycle. Use the words in the box.

| Sun rain clouds evaporation condensation lake underground water river sea mountain snow run off |

2 Write a short definition of the water cycle. Use the information that you know. Compare your definition with one in a dictionary.

_____

_____

_____

73

Topic 6 The Earth and Space

Student's Book p 88
6.5 Reservoirs and dams

Reservoirs and dams

Complete the table to show the positive and negative effects of reservoirs and dams

	Positive effects	Negative effects
Dams		
Reservoirs		

Topic 6 — The Earth and Space

Student's Book p 88
6.5 Reservoirs and dams

A dam or reservoir in my country

Do some research on a dam or reservoir in your own country. Answer the questions and record your research on this page.

1. What is the name of the dam or reservoir you researched?

2. Where is the dam or reservoir?

3. Why and when was it built?

4. What are the positive effects or benefits of the dam or reservoir?

5. Are there any negative effects?

Topic 6 The Earth and Space

Student's Book p 90
6.6 The Earth orbits the Sun

The Earth rotates (1)

Read the statements below.
Tick (✓) to show if each statement is TRUE or FALSE.
Then correct the false statements.

1 The Earth takes 24 hours to make a rotation around the Sun.
True ☐ False ☐

_____

2 The Earth rotates on its own axis. True ☐ False ☐

_____

3 The Sun moves around the Earth. True ☐ False ☐

_____

4 The Earth is flat. True ☐ False ☐

_____

5 We have day and night on Earth because of the Earth's rotation on its own axis. True ☐ False ☐

_____

6 The orbit of the Earth is the shape of a circle. True ☐ False ☐

_____

7 The Earth moves in an anticlockwise orbit around the Sun.
True ☐ False ☐

_____

Topic 6 The Earth and Space

Student's Book p 90
6.6 The Earth orbits the Sun

Thinking about evidence: data about the Sun

1 Study the data below.

Cape Town, South Africa				Tunisia, North Africa			
Date	Sunrise	Sunset	Length of day	Date	Sunrise	Sunset	Length of day
1 Feb	06:08	19:52	13h 43m	1 Feb	07:22	17:45	10h 22m
2 Feb	06:09	19:51	13h 42m	2 Feb	07:21	17:46	10h 24m
3 Feb	06:10	19:50	13h 40m	3 Feb	07:20	17:47	10h 26m
4 Feb	06:11	19:49	13h 38m	4 Feb	07:19	17:48	10h 28m
5 Feb	06:12	19:48	13h 36m	5 Feb	07:18	17:49	10h 30m
6 Feb	06:13	19:48	13h 34m	6 Feb	07:17	17:50	10h 32m
7 Feb	06:14	19:47	13h 33m	7 Feb	07:16	17:51	10h 34m
8 Feb	06:15	19:46	13h 31m	8 Feb	07:15	17:52	10h 36m
9 Feb	06:16	19:45	13h 29m	9 Feb	07:14	17:53	10h 38m
10 Feb	06:17	19:44	13h 27m	10 Feb	07:13	17:54	10h 40m

2 What does this data tell us about the position of the Sun?

_____

_____

3 Decide whether you will use a line graph or a bar chart to display the data.

4 Draw your line graph or bar chart to show how the number of hours of sunlight changes over this 10-day period in each of these countries.

continued

Topic 6 The Earth and Space

Topic 6 The Earth and Space

Student's Book p 90
6.6 The Earth orbits the Sun

The Earth rotates (2)

1 Draw a diagram to show the way the Earth turns on its axis. Label your diagram.

2 Read this statement and answer the questions.

The rotation of the Earth on its own axis causes day and night on Earth.

a What is an axis?

b What is rotation?

c Explain the way that rotation causes day and night.

Topic 6 The Earth and Space

Student's Book p 92
6.7 Why are there seasons on Earth?

Investigation report: What would happen if the Earth was not tilted?

Use your model of the Earth to find out what would happen if the axis of the Earth was not tilted. Then record your investigation here.

1 What equipment did you use?

2 What did you predict?

3 Explain briefly how you carried out your investigation.

4 What did you observe?

5 Did you check your observations? If so, what did you do?

6 Were your predictions correct?

7 What did your group conclude after this investigation?

Topic 6 — The Earth and Space

Student's Book p 92
6.7 Why are there seasons on Earth?

Why there are seasons on Earth

1 Draw a diagram to explain why the tilt of the Earth causes seasons on Earth.

2 What causes seasons on Earth?

3 Why do seasons occur at different times of the year in different places?

4 If it is summer in Australia, which is in the southern hemisphere, what season will it be in France, which is in the northern hemisphere?

Topic 6 The Earth and Space

Student's Book p 94
6.8 Satellites

What do you know about satellites?

1 Say if these statements are true or false. Correct the false statements.

　a The Earth is a natural satellite. True ☐ False ☐

　_____

　b A satellite is a small object that moves around in Space.
　True ☐ False ☐

　_____

　c The Moon takes about 28 days to move around the Earth.
　True ☐ False ☐

　_____

　d The Moon reflects light from the Earth. True ☐ False ☐

　_____

　e Some planets in our Solar System are not satellites. True ☐ False ☐

　_____

2 Complete the sentences.

　a The _____ space telescope is an example of an artificial satellite.

　b Artificial satellites _____ the Earth.

continued

Topic 6 The Earth and Space

3 Give two reasons why artificial satellites are useful to people on Earth.

_____

_____

4 Draw a diagram of the Moon and the Earth in Space. Label your diagram.

Appendix 1

Units for physical quantities

Quantity	Units
Length (used for length, height and width)	mm, cm, m, km
Area	cm², m²
Volume	ml, l, cm³, m³
Weight	N
Mass	g, kg
Time	s, min, h
Force	N
Gravity	N/kg
Temperature	°C
Sound	dB